WILDLY CUTE ANIMALS!

RED PANDAS

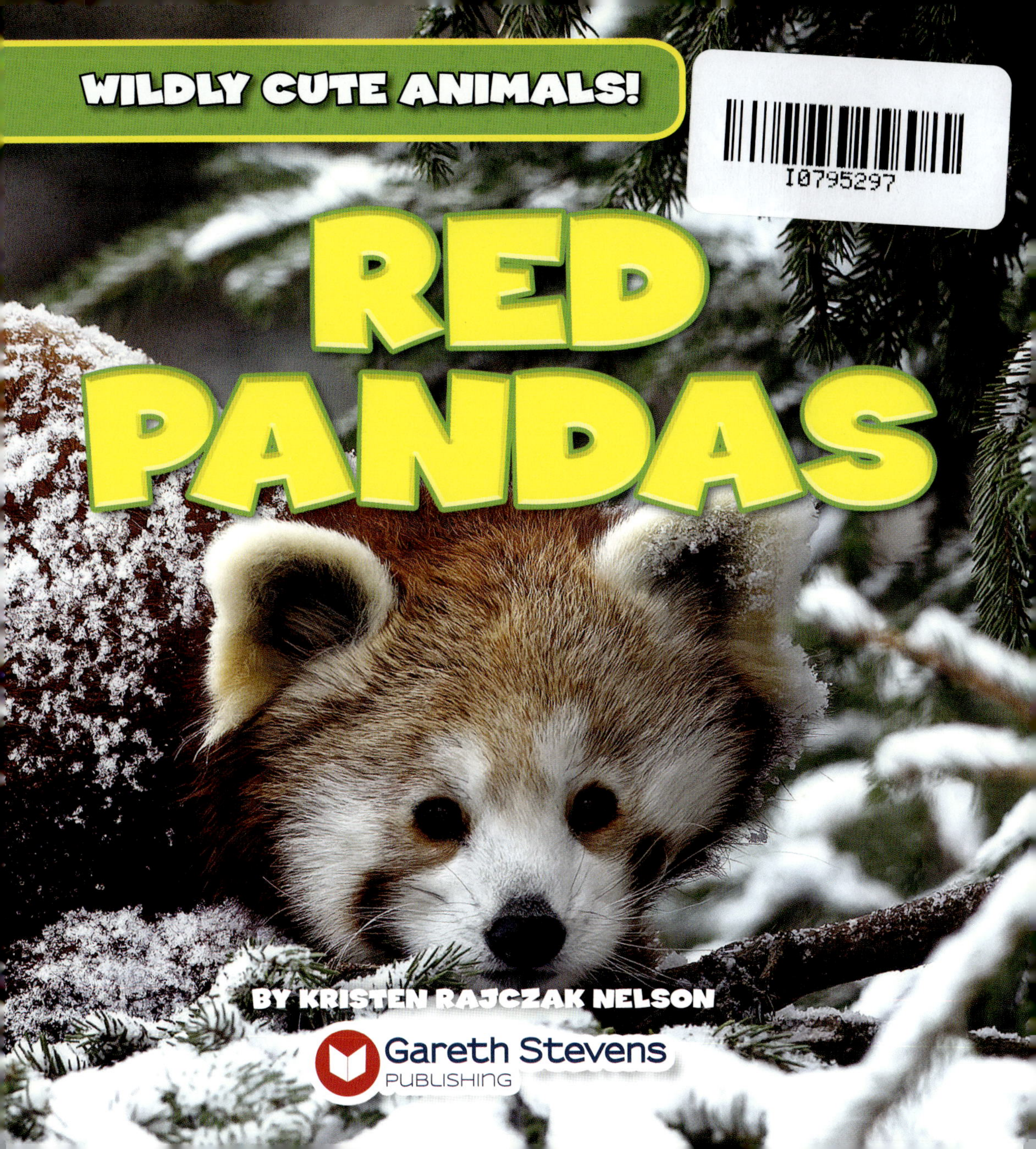

BY KRISTEN RAJCZAK NELSON

Gareth Stevens PUBLISHING

Please visit our website, www.garethstevens.com. For a free color catalog of all our high-quality books, call toll free 1-800-542-2595 or fax 1-877-542-2596.

Cataloging-in-Publication Data
Names: Rajczak Nelson, Kristen.
Title: Red pandas / Kristen Rajczak Nelson.
Description: Buffalo, NY : Gareth Stevens Publishing, 2026. | Series: Wildly cute animals! | Includes glossary and index
Identifiers: ISBN 9781482471236 (pbk.) | ISBN 9781482471243 (library bound) | ISBN 9781482471250 (ebook)
Subjects: LCSH: Red panda–Juvenile literature.
Classification: LCC QL737.C214 R35 2026 | DDC 599.76'3–dc23

Published in 2026 by
Gareth Stevens Publishing
2544 Clinton Street
Buffalo, NY 14224

Designer: Jeffrey Taylor
Editor: Kristen Nelson

Photo credits: Series art (tan background) Elena_Design/Shutterstock.com; Cover, p.1 Vladimir Jurek/Shutterstock.com; p. 5 Robrecht Kruft/Shutterstock.com; p. 7 shocky/Shutterstock.com; p. 9 vyasphoto/Shutterstock.com; p. 10 Satinder Pal/Shutterstock.com; p. 11 Martin Pelanek/Shutterstock.com; p. 13 JaysonPhotography/Shutterstock.com; p. 15 bachgui/Shutterstock.com; p. 17 slowmotiongli/Shutterstock.com; p. 19 gemadrun/Shutterstock.com; p. 21 Anton Watman/Shutterstock.com.

Printed in the United States of America

CONTENTS

Boldface words appear in the glossary.

Two of a Kind

Have you ever seen a red panda at the zoo? These small animals are wildly cute! Though they share part of a name, red pandas aren't **related** to giant pandas. In fact, they are in their own animal family! There are two kinds of red pandas.

The Red Panda Look

A red panda looks a bit like a kitten and a bit like a teddy bear! These **mammals** have reddish colored fur. Their belly fur is black. They have four legs ending in four paws. The paws have **claws** that help them climb.

Red pandas are about the size of a pet cat. They can stand as tall as 25 inches (63 cm). Their tail may be almost as long! The tail is bushy. Some red pandas have rings of darker fur on their tail.

Where in the Wild?

Red pandas live in the Asian countries of Nepal, India, Bhutan, China, and Myanmar (Burma). They live in the forests of the Himalayas. This is a mountain range that has the highest mountains in the world.

Red pandas spend a lot of time in trees. They lay in the sun and sleep in trees. Their coloring helps them **camouflage** among the reddish **moss** that grows on the trees where they live.

Bamboo Eaters

The forests red pandas live in must have lots of bamboo. That's because red pandas eat as much as 4 pounds (1.8 kg) of bamboo each day. Fruit, eggs, and birds may be part of their **diet** too. This makes red pandas omnivores, or eaters of both meat and plants.

BAMBOO

Living Solo

Red pandas like to live alone. When they do meet each other, red pandas **communicate**. They move their tails and heads in certain ways. They bark, squeal, and make other special sounds. These mammals only pair up to make babies.

Cute Cubs

Mother red pandas make a nest in tree roots or holes. They make nests soft using moss. Mothers commonly have about two babies at a time. Babies are called cubs. They stay with their mother for around a year after birth.

Endangered

Red pandas are endangered. This means they are in danger of dying out. Humans are cutting down the forests where they live. Red pandas get sicknesses that come from farm animals and pets. Sometimes, they are killed for their fur too.

GLOSSARY

camouflage: Colors or shapes on animals that allow them to blend with their surroundings.

claw: An animal's sharp nail.

communicate: To talk through sounds and movement.

diet: The food an animal eats.

mammal: A warm-blooded animal that has a backbone and hair, breathes air, and feeds milk to its young.

moss: A kind of plant that grows in clumps or mats in the shade.

related: Connected to as in a family.

FOR MORE INFORMATION

BOOKS

Bittel, Jason. *Mountain: Earth's Incredible Habitats*. New York, NY: Dorling Kindersley Publishing, Incorporated, 2024.

O'Brien, Cynthia. *Giant Panda vs. Red Panda*. Minneapolis, MN: Kaleidoscope Publishing, Inc., 2025.

WEBSITES

Red Panda | Species | WWF
worldwildlife.org/species/red-panda
Check out more about red pandas and the threats they face in the wild.

Red Pandas, Facts and Photos
nationalgeographic.com/animals/mammals/facts/red-panda
See cute pictures and read great facts about red pandas.

INDEX